ANCIENT
MEASUREMENT

ANCIENT MEASUREMENT

How Ancient civilizations created Precise
and Reproducible Standards

ROLAND A. BOUCHER

ARCHWAY
PUBLISHING

Archway Publishing books may be ordered through booksellers or by contacting:

Archway Publishing
1663 Liberty Drive
Bloomington, IN 47403
www.archwaypublishing.com
844-669-3957

ISBN: 978-1-4808-9534-8 (sc)
ISBN: 978-1-4808-9536-2 (hc)
ISBN: 978-1-4808-9535-5 (e)

Library of Congress Control Number: 2020917297

Print information available on the last page.

Archway Publishing rev. date: 12/02/2020

CONTENTS

ILLUSTRATIONS

TABLES

ACKNOWLEDGMENTS

This work was possible only because of the work of those who preceded me, without which this study would have been impossible. I speak of Dr. Marvin Adell Powell in his thesis at the University of Minnesota, Mr. A. E. Berriman O. B. E. in his 1953 book, *Historical Metrology*, and weights reported by Sir Arthur J. Evans in an article, "Minoan weights and Mediums of Currency from Crete, Mycenae and Cyprus," in *Corolla Numismatica*, published by Oxford University Press.

I would also like to express my appreciation to Julia M. Gelfand and Becky Imamoto, librarians at the University of California Irvine for their many helpful suggestions, and to Mr. Darcy Staggs, a fellow retired engineer who has reviewed my many PowerPoint presentations over these past five years.

And to Elizabeth Bray at the British Museum, Adele West at the Ashmolean Picture Library, and Gladys Pilastrini at the Musée du Louvre for invaluable assistance in obtaining photographic images used in our sixty-seven-year-old primary reference. Thanks to Elinor Wynne Lloyd at Its ALL Greek for permission to use the image of an amphora. And finally, to Monica Bryant, assistant curator in Nashville, Tennessee, thanks for permission to use the images of their magnificent replica of the Parthenon.

CHAPTER 1

INTRODUCTION

BACKGROUND

In the 1960s while working in the Hughes Aircraft Satellite Division, I was involved in demonstrating satellite communications to the airline industry when NASA began asking for reports in the metric system. It was not used in the aviation industry at that time.

This demand caused a number of us to study the history of the metric system. We found that in the sixteenth century, Queen Elizabeth I introduced the British Imperial System and about seventy years later, the French introduced their pendulum-derived metric system.

I will never forget the length of their new yard (meter). It was 993.7 millimeters. Fifty-five years later, when studying ancient Sumerian standards of measure, I found a yard that was exactly 993.7 millimeters long.

The obvious question was: did the Sumerians develop these standards using a pendulum? I had to find out. This report describes some of what was discovered.

HISTORY

In 1671 the Frenchman Jean Picard proposed a measurement system where a pendulum operated at 45 degrees north latitude with a beat (half period) of one second would establish the length of the new standard French yard to be called a meter. This meter would be subdivided into 100 centimeters and 1,000 millimeters. The standard of volume to be called a liter would be established as the volume of a 0.1-meter (10-centimeter) cube. This liter could be considered the volume of 1,000 cubic centimeters, or 0.001 the volume of 1 cubic meter. The standard of weight would be established as the weight of one liter of water at the temperature of maximum water density of approximately 4 degrees Celsius.

Five thousand years earlier, the Sumerians would develop exactly the same standard length to be called a step and the same standard volume to be called a sila. The weight of half-sila of water at room temperature would become the mina. When the French invented the metric system, it was already five thousand years old.

The Sumerian rules used to convert a standard of length to one of volume and weight were identical to the French proposal. They would be used for five thousand years until Queen Elizabeth I, chose to change the rules in the sixteenth century.

Proof that these ancient standards of length were pendulum-derived is not always as easy as comparing pendulum-derived length to an ancient standard of length. In some cases, the only standards that remain today are those of volume or weight. Fortunately, ancient standards of volume were derived directly from the cube of a linear dimension. Standards of weight, in turn, were derived from the weight of a standard volume of water at room temperature. In some cultures, standards of weight existed for a variety of grains as well; however, a standard using water was always established.

While there may be thousands of ancient weights in museums throughout the world, we were fortunate to have available a list of seventy-four weight-certified to be "standards of the land" by Dr. Marvin Adell Powell Jr. in his doctoral thesis at the University of Minnesota (1971). Later in our study, we would add weights certified by A. E. Berriman O.B E. in *Historical Metrology* (1953) and weights reported by Sir Arthur J. Evans in an article, "Minoan weights and mediums of currency from Crete, Mycenae and Cyprus," in *Corolla Numismatica*, published by Oxford University Press.

In order to establish the length of a pendulum as an official standard, the Sumerians would have been required to establish accurate and reproducible intervals of time. Fortunately, the Sumerians were expert astronomers and would have had no trouble establishing intervals of time based on the motion of the Moon, Sun, and stars, as well as the motion of the planet Venus.

The time interval for the full moon to travel one diameter in the night sky at apogee is a little over 121 seconds. This time interval was used to establish what I called the lunar standard of Lagash. The Sumerians used a sexagesimal system of mathematics and had divided the circle into 360 parts, which we call degrees. The time interval for the Sun to travel one degree in the sky is four minutes, or 240 seconds, which was used to develop the solar standard of Ur. A star appears to travel a little faster and the planet Venus a little slower. A star was used to develop the Egyptian foot while Venus was used to establish the Minoan foot.

We found that the Sumerians in Lagash had discovered that their cable of 360 steps, or yards, was a little longer than the length 1/360 of a degree on the polar circumference of the Earth. They made three attempts to adjust the length of their cable of 360 steps to be equal to 1/360 of a degree on the polar circumference. The length of this cable would establish the length of 1,000 new geodetic feet.

Table 1.1 summarizes the ancient standards of length and the resulting volumes and weights this report will establish.

#	Timing from	Step (mm)	Foot (mm)	Name of Standard
1	Moon	1011.24	337.1	Lunar standard of Lagash
2	Sun	993.30	331.1	Solar standard of Ur
3	Star	987.52	329.20	Assyrian foot of Babylon
4	Star	820.81	299.56	Egyptian foot, cubit, etc.
5	Venus	829.35	303.5	Minoan foot and pound
6	Sun	880.50	322.3	First geodetic foot of Lagash
7	Sun	882.20	317.60	Second geodetic foot of Lagash
8	Sun	853.47	307.3	Third geodetic foot of Lagash
9	Venus	857.09	308.5	Geodetic foot of Athens

TABLE 1.1. STANDARDS OF LENGTH (AS WELL AS VOLUME AND WEIGHT) TO BE ESTABLISHED BY THIS REPORT.

CHAPTER 2

THE ACCURACY WE COULD EXPECT FROM AN ANCIENT PENDULUM

The platinum ball and iron wire used by the French in their first proposed metric system would not have been available to the Sumerians. However, they would have gold, copper, or stone available for the ball and waxed flax string to replace the iron wire. I constructed several such pendulums using brass or steel balls and waxed flax string and found that a 944-millimeter pendulum would consistently swing through 100 beats in 100 seconds, a precision of one part in 10,000. The Sumerians could easily reproduce this pendulum.

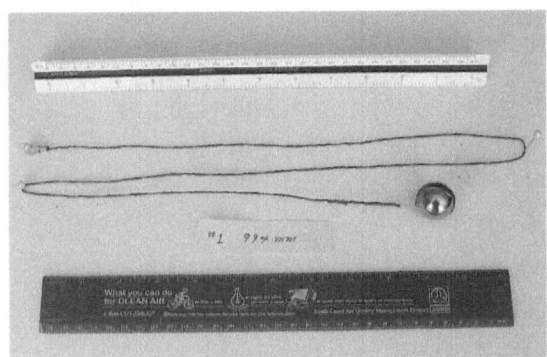

Figure 2.1. 994-millimeter test pendulum.

HOW A PENDULUM CAN BE USED TO ESTABLISH A UNIT OF LENGTH

Developing a comprehensive and precise set of measurement standards was not a trivial problem faced by ancient civilizations. Human body parts are not precisely reproduced from parent to child and are not suitable as standards of measurement. The names of some body parts have provided convenient names for some standards, however. Nature provides us no reliable standard for length, weight, or volume. Fortunately, nature can provide us at least four standards of time:

- full Moon rising or setting or passing by a vertical line of sight
- line of sight to the Sun rotating one or more degree in azimuth
- line of sight to a star rotating one or more degree in azimuth
- line of sight to Venus rotating one or more degree in azimuth

The "beat" of a pendulum is the time it takes the pendulum to make one-half swing, that is, the time from when the swinging pendulum is vertical or the point where it would hang if at rest, through the extension of the swing to its highest point until it swings back and reaches vertical again.

A PENDULUM CAN CONVERT ANY FRACTION OF A DAY INTO PRECISE UNITS OF LENGTH

The beat is determined almost exclusively by the Earth's gravity and the length from the pendulum's pivot point to the center of its mass. Unless the physical properties of a standard pendulum are carefully chosen, errors can occur.

Increasing the angle of the swing will cause the period of a pendulum to increase, while increasing the weight of the string relative to the ball will cause the period to decrease. These compensating effects can completely cancel each other, and with care, five-figure accuracy can be achieved.

L: length of string

M: mass attached to string

Pivot: support point of string

Alpha: maximum angle of swing

The length of a pendulum is proportional to the square of the period of swing. The period for 100 beats of a 1-meter simple pendulum is 100.384 seconds. A simple pendulum is one where the string has no weight and the ball is a point mass.

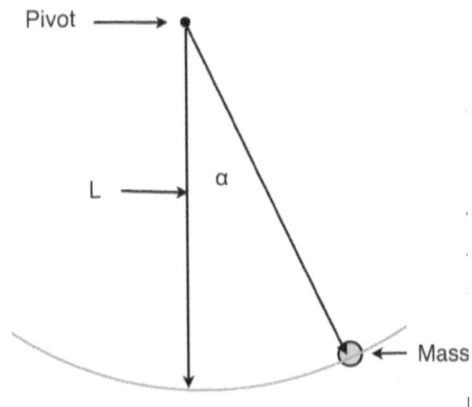

FIGURE 2.2. A SIMPLE PENDULUM.

A more detailed analysis of the pendulum can be found in Appendix 4.

In any case, it would be highly unlikely to expect an error greater than 0.1 percent (1,000 parts per million [ppm]) even if the user had no knowledge of this effect.

I have used two methods to establish the true length of a pendulum, which is essential to the success of this method. Figure 2.3 shows both.

The first method, used with a single-string pendulum, made use of a triangular opening perpendicular to the length of the pendulum to be measured. This triangular opening would center the ball for an accurate measurement. The second method was to pierce the ball on its center and support it with two strings to the pivot point. When measuring the pendulum, one would simply pull the string out to double length and divide the measured length in half.

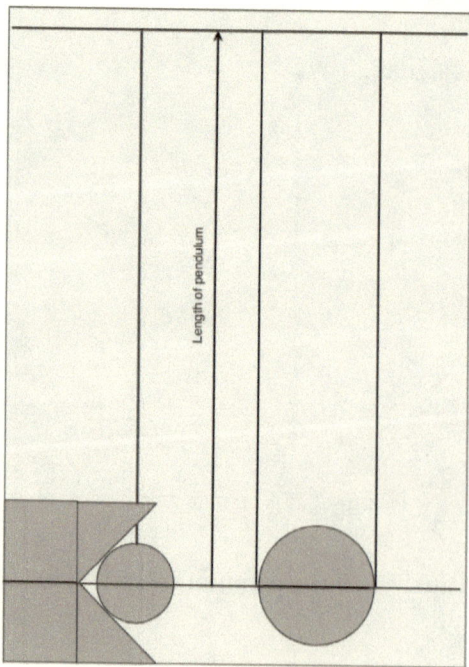

FIGURE 2.3. MEASURING THE LENGTH OF A PENDULUM.

The ancient Sumerians would have no trouble duplicating either method and may well have devised a superior one. Our only remaining task was to develop a simple and logical method to produce an accurate measure of time.

CHAPTER 3

ESTABLISHING ACCURATE INTERVALS OF TIME

The simplest method that seems to have been used was to mark the interval of time it took the diameter of a full moon to rise or set or the interval of time it took the moon's disk to pass a north-south line of sight. This interval, when viewing the full moon near apogee, is about 121 seconds. The length of a simple pendulum completing 60 swings, or 120 beats, in this interval would be about 1,012 millimeters.

The second method, which seems to have been used, was to apply the interval of time it took the sun's shadow or image to rotate through a fixed angle west to east. If the sun is low on the horizon, a small angle can be constructed by using a wheel to mark a distance of 180 diameters in a north-south direction from a peg. Then roll the wheel for a half circumference to each side for each degree of the south end of the line, planting stakes at both ends and stretching a string from the peg to each stake. The interval of time it takes for the center of the sun's shadow to move from one string to the next is 240 seconds per degree. Much better accuracy could be achieved by timing over a 60-degree angle or even over a full 24-hour day.

The Egyptians may have been the first to use a star to mark an interval of time. The interval of time it takes for a star in the equatorial plane to move 1/366 of a complete circle is 235.421 seconds. A star, as a mere pinpoint of light, can provide a much higher level of precision than the Sun. The length of a simple pendulum allowed to complete 366 beats in this interval would be about 300 millimeters.

The Minoans on the island of Crete may have been the first to use the planet Venus to mark the interval of time. The planet Venus is closer to the sun than the Earth and orbits the sun in 244 days. By viewing Venus when it is on the opposite side of the Sun from the Earth, its motion cancels out some of the apparent motion caused by the spinning Earth. The interval of time it takes for Venus to move 1/366 of a circle is 236.504 seconds. The length of a simple pendulum allowed to complete 366 beats in this interval would be about 303 millimeters. Creating an accurate interval of time of about four minutes, or multiples thereof, was obviously something the ancients would have no trouble achieving.

Element	Solar Day	Star Day	Venus Day
Length of day	86,400 seconds	86164.08 seconds	86560.33 seconds
Length of 1/360 day	240.00 seconds	239.345 seconds	240.4454 seconds
Length of 1/366 day	236.065 seconds	235.421 seconds	236.504 seconds

TABLE 3.1. TIMING INTERVALS FOR SUN, STAR, AND VENUS.

TIMING FROM THE MOON

Our task now is to convince the reader that the ancient Sumerians could have devised methods to accurately measure time using the motion of the Moon, Sun, stars, and planet Venus as their clock.

First let us examine the motion of the Moon in the night sky. Its orbital inclination is about 5 degrees, so at the 31-degree latitude of Lagash, the moon would rise to a maximum elevation above the horizon of between 54 and 64 degrees during its monthly cycle.

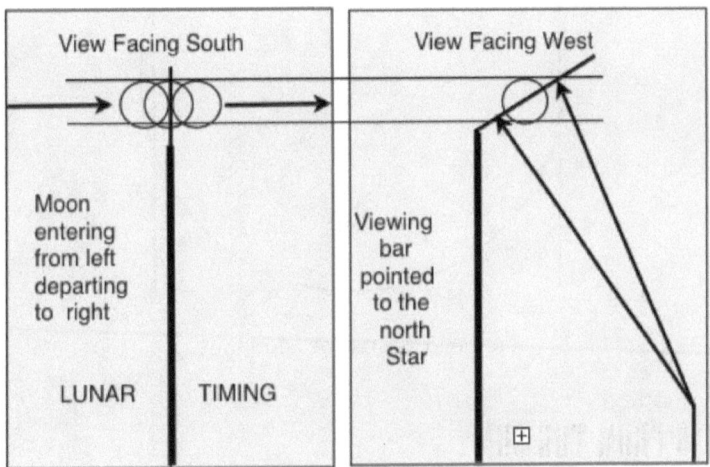

FIGURE 3.1. TIMING FROM THE MOON.

The time interval for the full Moon to cross the viewing bar at apogee is 121.2 seconds.

The Sumerians were excellent astronomers and would have had no trouble designing and building this apparatus. The apparent Moon is a little less than a half degree in diameter. It would be difficult to obtain much better than 1 percent accuracy unless timing took place over a number of apparent Moon diameters.

Figure 3.2 below depicts the Moon crossing the lunar timing bar on the right and leaving on the left. If the timing bar were 1 percent of the width of the apparent Moon diameter, it would be possible to measure a reduced time interval by twice the width of the bar, or 2 percent short. If a tiny gap were allowed, the measure would be a little long.

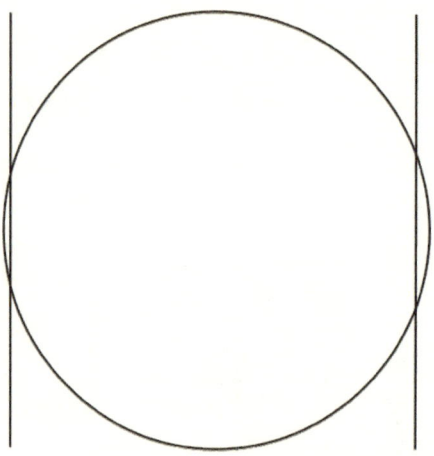

FIGURE 3.2. LOSS OF ACCURACY.

TIMING FROM THE SUN

One could construct a solar observatory by establishing a small hole in the wall or roof of a temple, allowing the Sun to create an image on the floor or wall. The motion of this image could be easily timed. For example, timing over 60 degrees would divide the day into six parts, or one could time for a full 24-hour day. The following sketches show how this could take place.

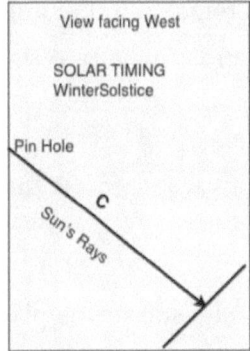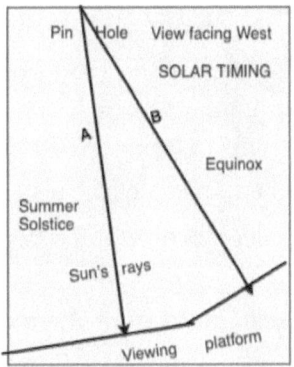

FIGURE 3.3. SUNRAYS
LOOKING WEST.

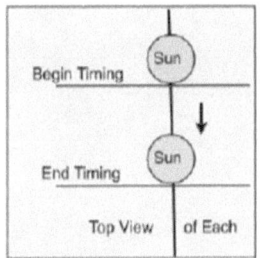

FIGURE 3.4. SUN IMAGE
SEEN FROM ABOVE.

Timing over 60 degrees would give us an acceptable error of about 1/1200. If the pendulum were timed for a full 360-degree day, the error would be much smaller at about 1/7200. Providing accurate timing from the Sun or Moon is not a trivial problem.

TIMING FROM A STAR OR THE PLANET VENUS

The stars arrive at the same position in the sky about four modern minutes earlier each day due to the Earth's orbital motion, which may be why the Sumerians divided the day into four-minute intervals they called a gesh. The star field appears to rotate 366 times in a year, so the number 366 was very important to an astronomer. A star is a mere pinpoint of light.

The apparent diameter of the planet Venus at apogee is 9.9 arc seconds, or 0.00275 degrees or 1/363.6 degrees.

Timing a pendulum from the motion of a star or the planet Venus can easily provide three - or four-figure accuracy.

CHAPTER 4

PENDULUM 1 AND THE LUNAR STANDARDS OF LAGASH

Pendulum 1 was timed with the Moon. If its length were adjusted to swing through 120 beats in 121.2 seconds, the time it took the Moon to travel one diameter in the night sky, its length as a simple pendulum would be 1,012 millimeters. The proof of its existence can be found in both a preserved standard of length and a preserved standard of volume as follows:

Berriman[5] states that Gudea was the governor of Lagash circa 2175 BCE. In 1881, de Sarzec found eight headless statues of Gudea in the ruins of Lagash, a port city in Sumeria. Two of the statues show Gudea with a ruler on his lap. The ruler had a scale of sixteen nominally equal divisions with a total length of 269 millimeters. The average length of the divisions is 16.81 millimeters.

If this length were a Sumerian *shusi*, then the length of Gudea's Sumerian *step* of 60 *shusi* (two cubits) would be 1008.75 millimeters. This length is only 0.4 percent shorter than our simple pendulum and is well within the expected range of a real pendulum. This would also establish Gudea's foot at *1/3 step*, *20 shusi*, or 336.25 millimeters.

FIGURE 4.1. GUDEA'S RULE.

FIGURE 4.2. ENTEMENA'S VASE.

Berriman[5] writes that Entemena's vase, a fine example of the silversmith's art (2400 BCE), was found by de Sarzec during his excavation of ancient Lagash at Tello. It is now in the Louvre. An inscription records its dedication by Entemena to the god Ningirsu in his temple of Eninnu, during Dudu's high priesthood. Entemena was the fifth governor at Lagash, during the Third Dynasty of Kish. Thureau-Dangin[5] published its volume as 4.71 liters. This is the volume of a 16.7626-centimeter cube.

If this volume were 1,000-cubic *shusi*, or the volume of a 10-*shusi* cube, the length of the 60-*shusi seed cubit* would be 1005.75 millimeters. The volume of Entemena's vase might also be considered a gallon of 1/8 cubic foot. The corresponding foot would be 335.25 millimeters. The length of a lunar pendulum would be adjusted to swing through 120 beats in 121.2 seconds.

P1 The Lunar Pendulum	Length (mm)	*Sila* (ml)	*Mina* (gm)	Dr. Powell and Mr. Berriman Matching Values
g = 9.80665	993.62	980.985	489.043	One-second beat period, standard gravity
g = 9.79487	992.43	977.452	487.282	One-second beat Lagash at 31.4068 degrees north
P = 1.009861	1012.1	1036.73	516.832	simple lunar pendulum string mass =0 ball diameter =0
C1 = -0.054 percent	1011.553	1035.062	516.000	516 grams #15 Susa 5 *shekels* Powell[7]
C2 = -0.054 percent	1011.553	1035.062	516.000	516 grams #16 Susa 5 *shekels* Powell[7]
C3 = -0.131 percent	1010.774	1032.580	514.763	514.8 grams #17 1/4 *shekel* Powell[7]
Correction Ave = - 0.08 percent	10011	1034.3	515.58	Ball diameter 2 fingers ball/string ratio = 57.1 1/2 swing = 10 fingers
C4 = -3.31 percent	1008.8	1026.48	511.723	1008.75-millimeter Gudea's rule[4]
Correction Use = 3.31 percent	1008.8	N/A	N/A	Ball diameter 2 fingers ball/string ratio = 33 1/2 swing = 15 fingers
L = 3 Feet	336.25	38018	37905	Foot, cubic foot, and *talent*
C5 = -1.62 percent	168.13	4725.25	N/A	4710-milliliter Entemena's vase = 1/8 cubic foot
C6 = -2.89 percent	335.28	N/A	N/A	335.28-millimeter Anglo-Saxon foot = L/3

TABLE 4.1. THE LENGTHS, VOLUMES, AND WEIGHTS FOR LUNAR PENDULUM P1.

In Table 4.1, we establish the theoretical length, volume, and weights, which would be developed from this simple pendulum, and then apply modest corrections for the length of a real pendulum. This table shows the summary of the results of quite a number of calculations and comparisons to certified weights.

In the first three rows, we compute the volume and resulting weight of a cube of water 0.1 the pendulum length on edge. The first row shows the results for a one-second pendulum at standard gravity as defined by the original French meter introduced in the year 1670. The second line corrects these calculations for operation at the latitude and gravity in the city of Lagash. It corrects the values for a pendulum with a little longer beat period resulting from lunar timing. Next, we compare these results with three certified weights from Dr. Powell's collection.[8] The average match is better than 0.1 percent in pendulum length and the characteristics of a real pendulum, providing a perfect match as shown in the right-hand column. In the last three rows, we introduce the foot of one-third pendulum length (yard), compare the results with Entemena's vase, and show a surprising match to the Anglo-Saxon foot.

It appears that Gudea's foot traveled to Europe where it became the Anglo-Saxon foot of 335.28 millimeters. This four-figure match in dimension is unlikely to have been the result of chance. This Anglo-Saxon foot then traveled to England, where the furlong of 600 Anglo-Saxon feet was used to establish all land boundaries. This furlong later became the British furlong of 660 British imperial feet, from which all British linear measures were derived.

The Sumerian city-states used what we might call a metric system, dividing their meter or yard by ten to generate a standard volume called a *sila* with *1/2 sila* of water being their standard of weight. This yard was divided into 60 *fingers* with 1/3 yard, or 20 *fingers*, making a Sumerian foot.

In table 4.1, we showed a talent created from the weight of one cubic foot of water. Sometimes the *talent* would be created from 60 *mina* rather than from a cubic foot of water. We hope this does not confuse the reader, but for completeness, we will show just such a talent in Table 4.2.

Example	Ratio	Weight (gm)	Matching Values
talent	60	30,905	30,900 *talent* #10, p. 358, Sir Arthur Evans Crete
mina	1	515.08	516 grams from Susa 2 *shekels* #16a Powell[7]
mina	1	515.08	516 grams from Susa 5 *shekels* #16b Powell[7]
60 shekels	1	8.58	8.58 = 4.29 x 2.25 *shekels* #17 Powell[7]

TABLE 4.2. TALENT OF 60 MINA AND SHEKELS.

CHAPTER 5

PENDULUM 2 AND THE SOLAR STANDARD STEP (METER) OF UR

When the French proposed their first metric system in the seventeenth century, they were unaware that it was already over five thousand years old and memorialized in the *mina* N. The original French proposal for a metric system in the early eighteenth century defined the meter as the length of a one-second pendulum (993.7 millimeters) when measured in the Earth's gravitational field at 45 degrees north latitude.

Pendulum 2 beat 240 times in 240 seconds and was timed with the Sun. The proof of its existence is found preserved in several ancient standards of weight.

FIGURE 5.1. *MINA* N.   FIGURE 5.2. LIMESTONE DUCK.

In the British Museum, there is a weight (#91005) that Berriman[6] calls "*mina* N" because its inscription certifies it to be a copy of a weight that Nebuchadnezzar II (605–562 BCE) made matching a weight that belonged to Shulgi of the Third Dynasty of Ur (c. 2100 BCE). Its mass weighed by Belaiew was 978.3 grams. This weight is equal to 981.1 milliliters of water at room temperature (25 degrees Celsius). This is the volume of a sila created from a 993.7-millimeter Sumerian step or double cubit.

In the Ashmolean Museum, there is a Babylonian limestone duck weight from Erech (Photo 1912.1162 Berriman[6]). Its published mass is 2,417 grams. If intended to be 5 *mina* in mass, one *mina* would equal 483.4 grams. A *talent* of 60 *mina* would weigh 29,004 grams. Its volume of water at room temperature of 20 degrees Celsius would be 29,090 milliliters. This is the volume of a 307.55-millimeter cube.

A double cubit at three times this length would be 992.65 millimeters, a little shorter than our simple pendulum. We conclude that the weight of the Babylonian limestone duck is derived from our simple one-second pendulum.

The length of pendulum 2, a one-second pendulum, would be adjusted to swing 240 beats in 240 seconds, or 1/360 of a solar day.

Pendulum 2	Length (mm)	Sila (ml)	Mina (gm)	Dr. Powell's[7] and Mr. Berriman's[10] Matching Values
g = 9.80665	993.621	980.985	489.043	One-second beat, standard gravity
g = 9.794032	992.343	977.204	487.157	One-second beat in Ur at string mass =0 ball diameter =0
CL1 = 0.003 percent	992.372	977.289	487.200	487.2 grams #52 5 *shekels* Powell[7]
CL2 = 0.05 percent	992.348	981.2	489.149	489.15 grams #50 2 *mina* N Powell[7]
CL3 = -0.058 percent	991.760	975.484	486.300	486.3 grams #53 30 *mina* Powell[7]
CL4 = -0.079 percent	991.556	974.882	486.000	486 grams #54 5 *shekels* Powell[7]
Correction Ave. = -0.032 percent	993.303	980.04	488.573	Ball diameter 2 fingers ball/string ratio = 48 1/2 swing = 10 fingers
CL5 = -0.026 percent	989.785	969.666	483.399	483.4 grams 2417 grams limestone duck[10]
CL6 = -0.030 percent	989.341	968.362	487.157	482.75 1.931 kilograms Assyrian lion weight #5
Correction Ave = -0.028 percent	989.564	969.018	483.078	Ball diameter 2 fingers ball/string ratio = 202 1/2 swing = 10 fingers
g = 9.8061999	993.576	980.852	997.952	993.7 millimeter French meter at 45 degrees north
CL7 = 0.0126 percent	993.701	N/A	N/A	Ball diameter 2 fingers ball/string ratio = 110 1/2 swing = 6 fingers

TABLE 5.1. THE LENGTHS, VOLUME, AND WEIGHTS FOR P2, A ONE-SECOND PENDULUM.

Table 5.1 shows the summary of the results of several calculations and comparisons to certified weights. In the first two rows, we compute the volume and resulting weight of a cube of water 0.1 the pendulum length on edge. The first row shows the results for a one-second pendulum at standard gravity as defined by the original French meter introduced in the year 1670. The second line corrects these calculations for operation at the latitude and gravity in the city of Ur.

Next, we compare these results with three certified weights from Dr. Powell's collection. The average match is better than 0.1 percent in pendulum length. The characteristics of a real pendulum, providing an average perfect match is shown in row 7 in the right-hand column. The next three rows show the match to the Assyrian limestone duck and lion weights, including the characteristics of a real pendulum, providing an average perfect match. The last two rows show the French metric pendulum first proposed in 1670 again, including the characteristics of a real pendulum, providing an average perfect match.

FIGURE 5:3. MINA D.

Mina D is the oldest extant weight in the Ashmolean Museum at Oxford, England. Dudu, the high priest at Lagash, signed it circa 2400 BCE. Berriman[6] reports that it was measured at 680.485 grams, exactly 150 Sumerian and 100 Minoan gold standards as well as 50 Egyptian Old Kingdom *deben*. *Mina* D and seven other gold standards are exact multiples of the weight of one cubic *finger* of water.

As mentioned in chapter 4, the Sumerians sometimes used the cubic foot as their standard. Some divided this cubic foot in a binary fashion, establishing what today we would call a bushel, gallon, and pint. In table 5.2, we establish these volumes as well as the cubic *finger* and *mina* D.

P 2	Length (mm)	Volume (ml)	Ratio finger	Weight (gm)	Matching Values
Bushel (cubic foot)	331.223	36338.2	8000	36231	Use French foot = 1/3 x 993.67 millimeters
Gallon	165.612	4542.6	1000	4528.8	no match found
Pint	82.806	567.83	125	566.105	no match found
cubic finger	16.561	4.5423	1	4.5288	4.53657 grams 1/150 *mina* D (below)
2/3 *Mina* D	N/A	4.55001	100	453.656	680.485 grams *mina* D #2 Powell[7]

TABLE 5.2. THE CUBIC FOOT, BUSHEL, GALLON, PINT, AND CUBIC *FINGER*.

In table 4.1, we showed a *talent* created from the weight of one cubic foot of water. Sometimes the *talent* would be created from 60 *mina* rather than from a cubic foot of water. We hope this does not confuse the reader, but for completeness, we show just such a *talent* in Table 5.3.

Pendulum 2	R	Weight	Matching values Weight
talent	60	29349	29,400 Arthur Evans *talent* at Knossos[8]
mina	1	489.154	489.154 *mina* N #50 Powell[7]
60 shekels	1	8.153	7.95 2/3 *shekel* #65 Powell[7]

TABLE 5.3 *MINA* N AND A CORRESPONDING TALENT AND SHEKEL.

In chapter 10, we will show that the Sumerians in the city of Lagash attempted to create a standard of measure based on the polar circumference of the Earth. Their goal was to establish the length of a pendulum or *step* such that the length of their *cable* of 360 *steps* would be equal to 1/360 of a degree on the polar circumference of the Earth. The length of this *cable* would establish 1,000 geodetic feet.

The length of this new foot would be 0.36 pendulum lengths rather than the customary one third. While we were unable to identify where this formula was misapplied in Sumeria, it seems to have traveled to foreign countries, where it was misapplied to the solar standard of Ur. We found two good matches in France and China, as shown in table 5.4.

Pendulum 2	Length (mm)	Sila (ml)	Mina (gm)	Matching Values
+ 126 ppm	993.576	N/A	N/A	993.7 millimeters French meter 1670
Foot at 0.36 L	357.72	N/A	N/A	360 millimeters Zhou Royal Ch ih[13]
Foot at 0.36 L	357.72	N/A	N/A	357.2 millimeters Bordeaux, France[14]

TABLE 5.4. TWO EXAMPLES WHERE THE FOOT IS 0.36 PENDULUM LENGTHS.

CHAPTER 6

PENDULUM 3, THE ASSYRIAN FOOT OF BABYLON, C. 1750 BCE

The Egyptian method of timing a pendulum with a star was later adopted by the Assyrians in Babylon. The original Sumerian one-second pendulum was allowed to swing the same 240 beats, but in 239.3447 seconds, or 1/360 of a celestial day. In table 6.1, we show that this pendulum appears to have created the Assyrian foot and provided a match to four signed Powell weights as well as the new Lion Weights.

Pendulum 3 (period)	Step (mm)	Sila (ml)	Mina (gm)	Measured Values
g = 9.80665	993.621	980.985	489.043	One-second beat period, standard gravity
g = 9.795322	992.473	977.588	487.523	One-second beat in Babylon at 32.5352 degrees north
P = 0.99727	987.061	961.685	479.421	**simple Assyrian pendulum** string mass =0 ball diameter =0
CL1 = 0.115 percent	988.196	965.005	481.076	481.07 #56 1/2 *mina* Powell[7]
CL2 = 0.05 percent	987.554	963.126	480.139	480.145 #57 1 *mina* Powell[7]
CL3 = 0.012 percent	987.179	962.029	479.592	479.6 #58 5 *shekels* Powell[7]
CL4 = 0.012 percent	987.179	962.029	479.592	479.6 #61 5 *shekels* Powell[7]
Correction Ave. = 0.047 percent	987.525	963.040	481.096	Ball diameter 2 fingers ball/string ratio = 89 1/2 Swing = 6 fingers
C6 = 0.134 percent	985.738	957.822	477.495	477.5 grams 2.865 kilograms Assyrian lion weight #3
C7 = 0.134 percent	985.738	957.822	477.495	477.5 grams 955 grams Assyrian lion weight #6
C8 = 0.006 percent	987.002	N/A	N/A	The Assyrian foot = L/3 = 329 millimeters

TABLE 6.1. THE ASSYRIAN *STEP, SILA, AND MINA*, ALONG WITH THE ASSYRIAN FOOT AT L/3.

In table 6.2, we show the cubic foot used to establish the *talent* weight and subdivided in a binary fashion by eight and sixty-four to establish the gallon, pint, and pound. Also, the weight of one cubic foot of grain was used to establish the *mina* of grain.

The Sumerian city-states sometimes created *the talent* weight from sixty *mina*; however, it was also created from the weight of one cubic foot of water.

Name	Ratio	Length (mm)	Volume (ml)	Weight (gm)	Measured Values
Foot = L/3	1.00	329.02	35618	35513	Values for simple Assyrian foot at L/3
Foot = L/3	1.00	329.00	35611	35506	Measured value for Assyrian foot
Foot = L/3	1.00	329.00	35611	35506	Cubic foot of volume and = *talent* weight
1/2 foot	0.50	164.50	4451.4	4438.3	1/8 cubic foot = or 1,000 cubic fingers
1/4 foot	0.25	82.25	556.40	554.78	1/64 cubic foot = pint or 125 cubic fingers
C = 0.062 percent	0.25	82.301	557.462	558.814	557.81 gm = pint 1/6 *mina* #7 Powell[7]
C = 0.062 percent	0.25	82.301	557.462	558.814	557.81 grams = pint 1/6 *mina* #8 Powell[7]
Grain *mina*	T/60	N/A	593.52	474.82	475 grams at 0.8 density 3 *mina* #67 Powell[7]
Grain *mina*	T/60	N/A	593.52	474.82	474 grams Assyrian lion weight #11

TABLE 6.2. THE FOOT, CUBIC FOOT, TALENT, GALLON, PINT, AND MINA OF GRAIN.

NOTE

The *talent* is the weight of one cubic foot of water (sixty-four pounds). The Troy pound is the weight of 1/60 cubic foot of grain (0.8 grams/milliliter).

Table 6.2 shows a summary of the results of several calculations and comparisons to certified weights. In the first three rows, we compute the volume and resulting weight of a cubic foot of water. In the next two, we establish the volume and weight of a half-and-quarter-size cube. Next, we compare these results with three certified weights from Dr. Powell's collection. The last two rows establish a *mina* of grain through one of Dr. Powell's weights and one Assyrian Lion Weight similar to the three shown below.

FIGURE 6.1. THE BEAUTIFUL ASSYRIAN LION WEIGHTS OF BABYLON.

CHAPTER 7

PENDULUM 4 AND THE EGYPTIAN STANDARDS

THE MEASUREMENT STANDARDS OF ANCIENT EGYPT DURING THE OLD KINGDOM

The Egyptian standards of length and volume were much more complex than most ancient standards. They were based on the five lengths: the *royal cubit, cubit, reman, foot,* and *finger* of 1/16 foot. The relative sizes are shown in Table 7.1.

Unit of Length	Relative Length	Relative Volume	Unit of Volume	Length (mm)	Volume (L)
Royal cubit	28	21952	*Deny*	524.148	144
Cubit	w24	13824	*Khar*	449.27	90.682
Reman	20	8000	*N/A*	374.39	52.478
Foot	16	4096	*N/A*	299.51	26.869
Finger	1	1	*N/A*	1.8719	6.56 (ml)
N/A	N/A	731.7	*Heqat = 1/30 Deny*	N/A	4.8
N/A	N/A	73.17	*Hinu = 1/300 Deny*	N/A	0.48
N/A	N/A	2.2867	*Ro = 1/9600 Deny*	N/A	15 (ml)

TABLE 7.1. RELATIVE AND APPROXIMATE SIZE OF EGYPTIAN LENGTHS AND VOLUMES.

We have developed two sets of standard lengths for the Egyptian Old Kingdom, both based on the length of a pendulum. The differences are small but profound, and both may have been used at different periods in the old kingdom.

The first is that developed by A. E. Berriman O.B.E. in *Historical Metrology* (London 1953). The second is based on our discovery that the circumference of the Great Pyramid is almost exactly three thousand Sumerian geodetic feet, a length developed by Sumerians in Lagash. This was their third attempt to establish a standard based on the measured polar circumference of the Earth. This was almost five thousand years before the French would attempt the same feat in their development of the metric system.

Mr. A. E. Berriman established the length of the royal cubit through the volume of bowls number twenty-seven and eight in the Petrie collection at the University College, London,[19] as follows:

- He assumed that bowls number twenty-seven and eight were each 1/16 and 1/24 *Deny* in volume.
- The volume of bowl number eight equals 1/24 *Deny*, which equals 1.25 *Heqat*. One *Deny* is 143.739 liters.
- The volume of bowl number twenty-seven equals 1/16 *Deny*, which equals 1.875 *Heqat*. One *Deny* is 143.006 liters.
- Alternately then, since the volume of the *Ro* equals 1 cubic *finger*, or 1/9600 *Khar*, the volume of bowl number eight is 400 *Ro*, and the volume of bowl number twenty-seven is 600 *Ro*.

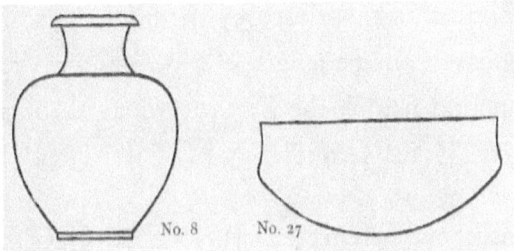

FIGURE 7.1. TWO EGYPTIAN BOWLS FROM THE PETRIE COLLECTION.

THE MEASURED VOLUMES

- Bowl number eight equals 366.2 cubic inches.
- Bowl number twenty-seven equals 546.5 cubic inches.
- Bowl number eight is 5989.14 milliliters.
- Bowl number twenty-seven is 8937.91 milliliters.
- One cubic inch equals 16.3548 cubic centimeters.

The length of the royal cubit and the foot based on the volume of each bowl are:

- Bowl number eight royal cubit equals 523.831 millimeters, and a foot equals 299.332 millimeters.
- Bowl number twenty-seven royal cubit equals 522.939 millimeters, and a foot equals 298.822 millimeters.

Mr. Berriman selected the length of the foot at 299.5 millimeters as a reasonable value. First, we will attempt to explain the origin of the lengths chosen by Mr. Berriman.

The Egyptians apparently realized that a star, a mere pinpoint of light, could provide a much higher level of precision than the Sun when measuring an interval of time as they developed their own standards in a manner similar to those used by the Sumerians. This standard appears to be based on the length of a pendulum, which beat 366 times in the period it took the Earth to rotate through 1/366 of a celestial day, or 235.421 seconds.

The Egyptians knew there were a little over 365 days in a solar year, and in a manner similar to the Sumerians, they created a *cable* of 365 *steps* to be equal in length to 1,000 feet. In table 7.2, we establish the resulting lengths of the *foot, finger, reman, cubit, and royal cubit.*

Pendulum 4	Length (mm)	Cable (m)	Matching Values
g = 9.80665	993.621	N/A	One-second beat, standard gravity
g = 9.7900640	991.94	N/A	One-second beat in Luxor at 25.6872 degrees north
P = 0.643237 seconds	820.812	N/A	(235.421sec)/366 in Luxor, Egypt
A cable of 365L	**820.812**	**299.56**	**The cable of 365L also equals 1,000 feet**
Foot of 16 fingers	820.812	299.56	***299.5 millimeters***, A. E. Berriman[19]
Reman of 20 fingers	820.812	374.36	374.4 millimeters, A. E. Berriman[19]
Cubit of 24 fingers	820.812	449.34	449.3 millimeters, A. E. Berriman[19]
Royal cubit = 28 fingers	820.812	524.23	524.2 millimeters, A. E. Berriman[19]
Finger of 1/16 foot	820.812	18.722	18.719 millimeters, A.E. Berriman[19]
Average correction 60 ppm	N/A	N/A	Ball diameter = 2 fingers Ball/string ratio = 125 1/2 Swing = 6 fingers

TABLE 7.2. EGYPTIAN STANDARDS BASED ON A PENDULUM THAT BEAT 366 TIMES IN 235.421 SECONDS WITH THE LENGTH OF A CABLE OF 365 STEPS = 1000 FEET.

Table 7.2 shows the results of several calculations. The first row shows the length for a one-second pendulum at standard gravity. The second line corrects these calculations for operation at the latitude and gravity in the city of Luxor. The third corrects the length for the period of the Egyptian timing method.

Next, we compare these results with those based on Mr. Berriman's bowl number eight. The average match is better than 0.1 percent in pendulum length. The characteristics of a real pendulum providing an average perfect match is shown in row ten in the right-hand column.

DERIVING THE EGYPTIAN STANDARDS FROM THE PERIMETER OF THE GREAT PYRAMID

The Great Pyramid of Giza was accurately measured in the nineteenth century by both Petrie and Cole[19], establishing the average width at 230.355 meters with a precision of better than one part in 10,000. The four sides are aligned north-south and east-west to within 1/15 degree of the true values. The perimeter of the Great Pyramid is 921.421 meters.

In chapter 12 when discussing the third geodetic standard of Lagash, we discovered that the Great Pyramid appeared to be designed to this standard with its perimeter equal to 3,000 Sumerian feet, or 1/120 of a degree on the polar circumference of the Earth.

Elements of Length	Cable (m)	Foot (mm)	Comments
Foot (Ave.)	N/A	299.072	Foot derived from average of bowls number eight and twenty-seven
Circumference of pyramid	921.589	307.196	3,000 Sumerian geodetic feet of Lagash
True circumference	921.421	307.140	Measured by Petrie and Cole in the nineteenth century
1760 royal cubits	921.421	523.535	If circumference = *1760 royal cubits*
Foot = 16 fingers	921.421	299.163	If circumference = 1760 *royal cubits*
Reman = 20 fingers	921.421	373.954	If circumference = 1760 *royal cubits*
Cubit = 24 fingers	921.421	448.744	If circumference = 1760 *royal cubits*
Royal cubit = 28 fingers	921.421	523.535	Cubic cubit = 143,495 cubic centimeters
1 pyramid finger	921.421	18.6977	Cubic finger = 6.5368 cubic centimeters

TABLE 7.3. EGYPTIAN STANDARDS DERIVED FROM THE
PERIMETER OF THE GREAT PYRAMID OF GIZA.

This geodetic pendulum design when operated in the gravity of Memphis would have produced a 3,000-foot perimeter of 921.589 meters. This value is only 184 parts per million longer than that established by Petrie and Cole. In table 7.3, we establish the length of the finger, foot, reman, cubit, and royal cubit.

Later in the New Kingdom, the royal cubit was eliminated, and the cubic cubit was renamed the Khar. In Table 7.4, we show standards of weight for both the New and Old Kingdoms.

Old Kingdom Name	Old Kingdom Weights	New Kingdom Name	New Kingdom Volume	New Kingdom Weight
Cubic Cubit	96 L (2/3 RC)	*Khar*	90.682 liters	90.414 kilograms
Deben (1)	13.61 grams	*Deben (1)*	N/A	91 grams
Sep (10)	136.1 grams	*Sep (10)*	N/A	910 grams
Kite (1/10)	1.361 grams	*Kite (1/10)*	N/A	9.1 grams

TABLE 7.4. EGYPTIAN STANDARDS OF WEIGHT
FOR BOTH OLD AND NEW KINGDOMS.

NOTE

The Old Kingdom deben equals three Sumerian or two Minoan gold standards. One Sumerian gold standard, the weight of one cubic finger of water, is 4.53656 grams. One Egyptian gold standard, the weight of three cubic *fingers* of water, is 13.6097 grams.

CHAPTER 8

PENDULUM 5, VENUS AND THE MINOAN STANDARD OF KNOSSOS

Venus was an important goddess to the Minoans (2700–1100 BCE). They timed their pendulum from Venus while in opposition for 366 beats during the time it took Venus to divide the rotation of the Earth by 366. The planet Venus is closer to the Sun than the Earth and orbits the Sun in 244 days. By viewing Venus when it is on the opposite side of the Sun, its motion cancels out some of the apparent motion caused by the spinning Earth, lengthening the period for 1/366 Venus day to 236.504 seconds. This essentially divided the celestial solar day into 365.25 parts. The length of the resulting *cable* was 303.6 meters, and the foot was 303.6 millimeters.

DETAIL OF CALCULATIONS REGARDING THE MINOAN FOOT

The Minoan pendulum beat 366 times in 236.504 seconds. The length of their cable of 366 pendulum lengths was equal to 1,000 Minoan feet. In Table 8.1, we establish the theoretical length for a simple pendulum and the resulting foot. Applying modest corrections for the period and length of a real pendulum results in the following *foot*, *sila*, and *mina* along with corresponding measured values from reliable sources.

Pendulum 6	Step (mm)	Sila (ml)	Mina (gram)	Measured Values
g = 9.80665	993.621	980.985	499.043	One-second beat at standard gravity
g= 9.7953220	992.473	977.588	487.532	One second Babylon at 33.5352 degrees north
g= 9.7975980	992.701	978.27	487.883	One second Knossos at 35.2985 degrees north
P = 0.646186	**829.02**	N/A	N/A	366 beats in 236.504 seconds
366 steps = 1,000 feet	**Foot**	**Cubic Foot**	**Talent**	The *talent* = weight of one cubic foot of water
Minoan *foot*	303.421	27934	27852	1000 Minoan *feet* = 366 steps(L)
C1 = 0.06 percent	303.600	27984	27901	303.6 millimeter foot A. E. Evans at Knossos[8]
C2 = 0.08 percent	303.640	N/A	N/A	303.64 millimeter early English foot[22]
C3 = -0.02 percent	303.360	27983	27900	27900 grams bronze *talent* #4 Halbherr[8]
Correction Ave = 0.04 percent	303.542	**N/A**	**N/A**	Ball diameter = 2 fingers Ball/string ratio = 92 1/2 Swing = 6 fingers

TABLE 8.1. MINOAN STANDARDS OF LENGTH, VOLUME, AND WEIGHT.

This table show a summary of the results of several calculations and comparisons to certified weights. The first row shows the results for a one-second pendulum at standard gravity. The second line corrects these calculations for operation at the latitude and gravity in the city of Babylon. The third line corrects these calculations for operation in the city of Knossos on Crete. The fourth row corrects the pendulum length for the reduced time interval used. The fifth and sixth row establishes the Minoan *foot*, cubic *foot*, and *talent* weight. Row eight establishes an almost-perfect match to an early English foot. Row nine, establishes an almost-perfect match to a bronze *talent* weight found on Crete. The tenth row shows the characteristics of a pendulum, which would produce a perfect match to the average lengths of line seven, eight, and nine.

The ancient Minoans divided their standards of volume (amphora) and weight(*talent*) of one cubic *foot*, either by sixty or by halves yielding volumes of 1/8 and 1/64 amphora. The standard volume for a pint of 1/64 amphora equals 437.25 cubic centimeters.

The standard weight for a *pound* was the weight of a pint of rainwater at 20 degrees Celsius, which equals 436 grams. The standard weight of a *Troy pound* was 1/60 amphora of wheat at 0.8 kilograms/liter, which equals 373.2 grams.

In table 8.2, we use the 27,900-gram bronze *talent* #4 in Table 8.1 to establish these subdivisions. We were quite surprised to find that the English Troy pound was of Minoan origin.

Pendulum 6	Ratio1/x	Weight(g)	Correction	Measured
talent	1	**27900**	N/A	27,900-gram bronze *talent* #4 Halbherr[8]
Pound	64	435.94	0.33 percent	437.4 grams English mercantile pound[6]
Cubic finger	40966	6.8115	0.03 percent	6.805 grams 1/150 *mina* D gold standard[7]
mina	60	465	None	465.004 grams #72 1/2 *mina* duck[7]
sila grain	60	372	0.33 percent	373.24 grams English Troy pound Zupko[26]

TABLE 8.2. THE MINOAN 29,700-GRAM TALENT DIVIDED INTO A POUND, GOLD STANDARD, *MINA*, ASSYRIAN *MINA*, AND *SILA* OF GRAIN OR ENGLISH TROY POUND.

THE MINOAN FOOT WAS IMMORTALIZED IN THE MAGNA CARTA

The accuracy of these measurements would suggest that the English mercantile pound and Scottish pound had been established precisely by the weight of one Minoan pint of rainwater. It also would appear that the Troy pound had been established precisely by the weight of 1/60 of a bushel (amphora) of wheat at 0.8 kilograms/liter. These English values were fixed by the Magna Carta of King John on June 15, 1215.

Moving to Japan, we find the Japanese *shaku* of 303 millimeters, a very close match to the Minoan foot when considering the difference in latitude. The Japanese, just as the Minoans, were a maritime nation. It is interesting that the largest linear standard of ancient Japan was the *ri* of 12,960 *shaku*, a length almost exactly that of 1/10,000 the polar circumference of the Earth.

CHAPTER 9

ESTABLISHING THE POLAR CIRCUMFERENCE OF THE EARTH AS A STANDARD OF LENGTH

The Sumerians in Lagash had discovered that their *cable* of 360 steps, or yards, was a little longer than the length 1/360 of a degree on the polar circumference of the earth. They made three attempts to create a **cable** of 360 steps, which was exactly 1/360 of a degree on the polar circumference of the Earth measured locally. The first overstated the circumference by about 5 percent and the second by about 3 percent, and finally the third geodetic standard was established with an error of less than 0.25 percent. Rather than use the new step as their geodetic standard, they chose instead to use a new foot where 360 of these new *cable* lengths would equal 1,000 geodetic feet. We don't know which measurement standard the Sumerians of Lagash used to measure the Earth. For convenience, we will use the modern meter for our calculations.

The latitude of Lagash is 31.4068 degrees north. To increase our baseline, we will assume that the most southern point of measurement would be the city of Ur at 30.9575 degrees north, and our most northerly point of measurement would be the city of Nineveh at 36.2296 degrees north. The difference in latitude of these two cities is 5.2721 degrees. Nineveh is not directly north of Ur, but 3.8656 degrees to the west, which will complicate our calculations. Converting these degrees to minutes of latitude and longitude, Nineveh is 316.336 minutes north and 231.936 minutes west of Ur. Nineveh is at 36.2 degrees north. One minute of longitude equals 1498.9 meters. And 3.8656 degrees west equals 347.65 kilometers.

Nineveh

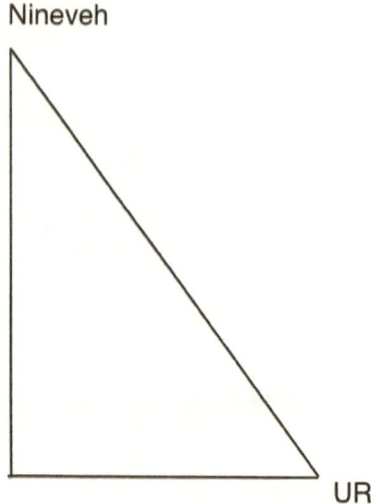

UR

The average latitude equals 33.6 degrees north. One minute of latitude equals 1848.59 meters. 5.2712 degrees equals 584.66 kilometers. The slant distance between the two cities is 680.212 kilometers. The angle in a table of tangents would be 59.26 degrees.

From the table, the sine of 59.26 degrees equals 0.8595. The distance north (680.212 x 0.8595) is 584.642 kilometers. The distance per degree equals 110.894 kilometers. The circumference (360 x 110.912 kilometers) is 39,921.7 kilometers. The actual circumference at 33.6 degrees north equals 39,929.5 kilometers.

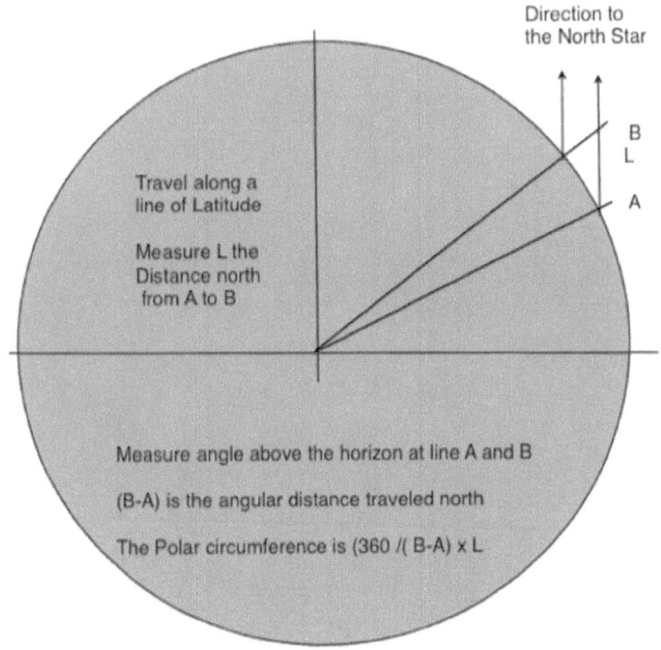

Direction to
the North Star

B
L

A

Travel along a
line of Latitude

Measure L the
Distance north
from A to B

Measure angle above the horizon at line A and B

(B-A) is the angular distance traveled north

The Polar circumference is (360 /(B-A) x L

FIGURE 9.1. MEASURING THE CIRCUMFERENCE OF THE EARTH.

CHAPTER 10

PENDULUM 6 AND THE FIRST GEODETIC FOOT OF LAGASH

The ancient Sumerians of Lagash recognized that the length of 360 of their *steps* (yards) was very nearly 1/360 of a degree on the polar circumference of the Earth. The old *step* was based on the length of a pendulum which beat 240 times in 240 seconds, 1/360 day or one Sumerian *gesh*. Its length derived from a simple pendulum was 993.7 millimeters. This new pendulum was much shorter. It would beat 360 times in 1/360 part of a day (240 seconds or one Sumerian *gesh*), and double its lengths would be used.

They declared a new geodetic foot, which was 1/1000 the length of a *cable* of 366 of these new double *steps*. It was intended to be 1/360 of a degree on the polar circumference of the Earth. The length of this first geodetic *cable* was 322.8 meters, establishing the geodetic foot at 322.8 millimeters. This resulted in a standard *mina* weight of about 522 grams. Proof of its existence can be found in Dr. Powell's weights number thirteen and fourteen.

Pendulum 6 1.5 beats/sec	Step (mm)	Sila (ml)	Talent (gm)	Mina (gm)	Measured Values
g = 9.80665	993.621	980.985	N/A	489.043	One-second beat period, standard gravity
g =9.794870	992.427	977.453	N/A	487.282	One-second beat Lagash at 31.4068 degrees
P = 0.66667	882.158	686.498	N/A	342.37	2/3 sec Pendulum in Lagash string mass =0 ball diameter =0
Foot = 0.366L	*Foot*	*Pint*	*Talent*	*Pound*	**First geodetic** *foot* **of Lagash**
Foot = 0.366L	322.870	525.9	33558	524.5	360 beats in 240 seconds
C = -0.159 percent	322.357	523..4	33398	521.85	522.0 matching weights #13 Powell[7]
C = -0.217 percent	322.17	525.9	33340	520.9	521.1 matching weights #14 Powell[7]
Average correction= -0.188 %	**322.263**	N/A	N/A	N/A	Ball diameter = 2 *fingers* Ball/string ratio = 64.7 1/2 Swing = 15 *fingers*
Surprise	**321.900**	N/A	N/A	N/A	**New value for Stonehenge long-foot**

TABLE 10.1. THE FIRST GEODETIC FOOT OF LAGASH AND MATCHING
WEIGHTS NUMBER THIRTEEN AND FOURTEEN FROM DR. POWELL.

Table 10.1 shows a summary of the results of several calculations and comparisons to certified weights. The first row shows the results for a one-second pendulum at standard gravity. The second row corrects these calculations for operation at the latitude and gravity in the city of Lagash. The third row corrects the pendulum length for the reduced time interval used. The fourth and fifth rows establish the first geodetic foot of Lagash, along with pint, pound, and *talent* weights. Rows six and seven show a match to two of Dr. Powell's certified weights. Row eight shows the characteristics of a pendulum, which would perfectly match the average of the certified weights.

A SURPRISE DISCOVERY

Line ten shows that this new Sumerian foot seems to have reached Scotland, where it became the Stonehenge long-foot. A recent article in the publication, *The British Journal for the History of Mathematics*, by Anne Teather, Andrew Chamberlain, and Mike Parker established a relatively accurate value of the Stonehenge long-foot at 321.9 millimeters based on the chalk drums of Fulton. The true value of the Stonehenge long-foot may not be precisely the 322.9-millimeter Sumerian foot or precisely the 322.2.9-millimeter value established by the chalk drums, but the search for the ancient roots and length of the long-foot of Stonehenge may at long last be within our grasp.

This first geodetic *cable* of 1,000 geodetic *feet* was about 5 percent longer than 1/360 of a degree on the polar circumference of the Earth measured at the latitude of Lagash. Two more efforts to find the true geodetic *foot* would be attempted.

CHAPTER 11

PENDULUM 7 AND THE SECOND GEODETIC FOOT OF LAGASH

The Sumerians in Lagash devised a second geodetic *foot*, which was 1/1000 the length of 360 *steps*. As in the first attempt, the pendulum beat 360 times in 240 seconds. A new cable of 360 of these new *steps* was now equal to 1,000 geodetic *feet*. The length of this new geodetic *foot* was 317.577 millimeters. The new foot was still somewhat long since the true value measured at the latitude of Lagash is about 308 millimeters. We searched for any evidence of the use of this geodetic foot in linear measurements in Mesopotamia but found none. However, when searching in *Historical Metrology* by A. E. Berriman, we found two, the 318-millimeter Chinese market foot of the Zhou Dynasty and a second match in the Fuss of Bern, Austria. These results are shown in Table 11.1.

Pendulum 7	Step (mm)	Sila (ml)	Mina (gm)	Matching Values
g = 9.80665	993.621	980.985	489.043	One-second beat period, standard gravity
g = 9.794870	992.427	977.453	487.282	One-second beat period Lagash at 31.4068 degrees
P = 0.66667	882.158	686.498	342.37	Two-thirds-second simple pendulum in Lagash
cable = 360 L	881.158	N/A	N/A	*cable* = 317.577 meters = 1,000 feet
foot = 0.360L	317.577	N/A	N/A	Second geodetic *foot* in Lagash
C = 0.133 percent	318.000	N/A	N/A	318-millimeter Chinese market foot[13] Zhou Dynasty
C = -0.182 percent	317.000	N/A	N.A	317-millimeter Fuss of Bern, Austria[15]

TABLE 11.1. THE EXPECTED VALUES FOR PENDULUM
7 WHEN MEASURED IN LAGASH.

This Table shows a summary of the results of several calculations and comparisons to certified weights. The first row shows the results for a one-second pendulum at standard gravity. The second row corrects these calculations for operation at the latitude and gravity in the city of Lagash. The third row corrects the pendulum length for the reduced time interval used. The fourth and fifth row introduce the second geodetic foot of Lagash. We found no Powell matches for the *mina* weights but did find two matches for the second geodetic *foot* one in China and a second in Bern, Austria.

When using the *foot* as a standard of length, it was common to use the cubic *foot* as a measure of volume with the weight of this cubic *foot* of water establishing the *talent*. This foot was usually divided into sixteen *fingers*. An eight-*finger* cube would later become a gallon in the West, and the four-*finger* cube with a volume of 1/64 cubic *foot*, becoming the pint and its weight the *mina*.

In Table 11.12, we establish proof of the existence of this second geodetic foot when we compare the weight of one cubic *foot* of water with that of *talent* number fifteen found by Sir Arthur Evans, and with the weight 1/64 cubic *foot* of water with Dr. Powell's certified weights in row seven, we show the characteristics of a pendulum, which produce a perfect match to the average of the four Powell weights shown. In row eight, we show a *mina* established by the weight of 1/60 *talent*.

Pendulum 7	Length (mm)	Volume (ml)	Pound (gm)	Matching Values (corrected 3/8/20)
1 cubic foot	317.577	32029	31934.6	32,000-gram *talent* #15 A. E. Evans
1/64 cubic foot	79.394	500.457	498.978	The weight of 1/64 *talent* as a *mina*
C1 = -0.020 percent	79.378	500.152	498.674	498.67 grams #31 Shulgi 10 *minas* Powell[7]
C2 = -0.034 percent	79.367	499.942	498.464	498.468 grams #32 Telloh 5 *shekels* Powell[7]
C3 = -0.065 percent	79.342	499.477	498.001	498.024 grams #33 Bronze lion Powell[7]
C4 = -0.065 percent	79.342	499.477	498.001	498 grams #34 5 *shekels* Powell[7]
Correction Ave = 0.046 percent	N/A	N/A	N/A	**Ball diameter = 2 fingers** **Ball/string ratio = 50** **1/2 Swing = 10 fingers**
1/60 *talent*	N/A	N/A	532.243	534.16 grams #12 duck 1/6 *mina* Powell[7]

TABLE 11.2. THE *TALENT* ESTABLISHING A *MINA* AND COMPARING IT TO MEASURED VALUES.

This second standard geodetic foot of 317.6 millimeters was still much larger than the true value of 308 millimeters when measured in Lagash. The Sumerians of Lagash would now replace it once again with a much more accurate value.

CHAPTER 12

PENDULUM 8, THE THIRD GEODETIC FOOT OF LAGASH AND PYRAMID OF GIZA

If the second geodetic pendulum of Lagash were allowed to beat 366 times rather than 360 times in 240 seconds, the length of the resulting *foot* would be 307.25 millimeters. This is within 0.24 percent of the true value of 307.99 millimeters. The length of one hundred of these new Sumerian *feet* is almost exactly one arc second on the polar circumference of the Earth. We were able to find two elements of conclusive proof of its existence, see Table 12.1.

Pendulum 8	Length (mm)	Sila (ml)	Mina (gm)	Measured Values
g = 9.80665	993.621	980.985	489.043	One-second beat period, standard gravity
g = 9.794870	992.427	997.454	487.282	One-second beat in Lagash at 31.4068 degrees
P = 0.655738	853.472	621.681	309.922	366/240 seconds in Lagash at 31.4068 degrees
foot = 0.360L	N/A	N/A	N/A	307.249 millimeters equals a geodetic foot in Lagash
g = 9.795322	992.47	977.589	487.35	One-second beat in Babylon at 32.5352 degrees north
P = 0.655738	853.512	621.768	309.965	366/240 seconds in Babylon at 32.5352 degrees
foot = 0.360L	N/A	N/A	N/A	307.264 millimeters equals geodetic foot in Babylon
New Header	*Foot*	*Cubic Foot*	*talent*	*talent* is the weight of one cubic foot of water
Foot = 0.360L	307.264	29009.2	28923.5	*The foot, cubic foot, and talent*
C = 0.088 percent	307.264	29009.2	28923.5	**2,900 grams, the octopus *talent* of Babylon**
C = 0.088 percent	N/A	N/A	**29000**	**Ball diameter = 2 fingers** **Ball/string ratio = 72** **1/2 Swing = 6 fingers**
E = -0.254 percent	307.264	N/A	N/A	**308.045 millimeter actual foot at 32.5352 degrees**

TABLE 12.1. THE THIRD GEODETIC FOOT OF LAGASH AND BABYLON ALONG WITH THE MAGNIFICENT OCTOPUS *TALENT* OF BABYLON.

In the first three rows, we compute the volume and resulting weight of a cube of water 0.1 the pendulum length on edge. The first two rows show the results for a one-second pendulum at standard gravity and then for the gravity in Lagash. The third row corrects for the timing actually used, and in the fourth row, we compute the length of the geodetic foot of Lagash. In rows five, six, and seven, we repeat the process for the city of Babylon.

Next, we introduce the *talent* as the weight of one cubic *foot* of water at room temperature and compare the result with the magnificent Octopus *talent* of Babylon. The last row compares the length of the geodetic *foot* of Babylon with modern satellite data.

Our first proof of the existence of the third geodetic *foot* was in the magnificent Octopus *talent* of Babylon, which was discovered in Knossos, Crete, in 1901 by Sir Arthur Evans. This magnificent 29,000-gram *talent* weight from circa 1650 BCE may well have been commissioned to celebrate the thousandth anniversary of the building of the Great Pyramid at Giza.

FIGURE 12.1. OCTOPUS *TALENT* WEIGHT.   FIGURE 12.2. OCTOPUS AMPHORA.

The second proof of the existence of the third geodetic *foot* of Lagash is in the surprising discovery that it was used in the design of the Great

Pyramid at Giza. The Great Pyramid of Giza was accurately measured in the nineteenth century by both Petrie and Cole, establishing its average width at 230.355 meters with a precision of better than one part in 10,000. The four sides are aligned north-south and east-west to within 1/15 degree of the true value.

FIGURE 12.3. THE GREAT PYRAMID OF GIZA.

Its perimeter was established as 3,000 Sumerian *feet*. Sometime before 2680 BCE, when the construction of the Great Pyramid began, the Egyptian astronomers and engineers would have become aware that a Sumerian geodetic pendulum provided an accurate measurement of the length of an arc-minute of latitude. This Sumerian pendulum design, when operated in the gravity of Memphis, would have produced a 3,000-*foot* perimeter of 921.589 meters. This value is only 182 parts per million longer than the pyramid as constructed.

Table 12.2 shows some of the details. The Great Pyramid of Giza as well as the length of the Egyptian royal cubit were both based on the polar circumference of the Earth as defined by the third geodetic *foot* of Lagash.

Pendulum 8	Lengths (mm)	Foot (mm)	Description
g = 9.794870	L = 853.472	307.249	Pendulum 8 in Lagash at 31.4068 degrees north
g = 9.793163	L = 853.323	307.196	Pendulum 8 in Memphis at 29.8300 degrees north
3000 *feet*	921.589 M	307.195	3,000 *feet* in Memphis, Egypt
			The measured value is 182 ppm smaller
1 / 3000 Perimeter	921.421 M	307.140	Pyramid measured by Petrie and Cole
1 / 1760 Perimeter	921.421 M	523.535	**Length of Egyptian royal cubit**
1 / 3080 Perimeter	921.421 M	299.163	**Length of Egyptian standard foot**

TABLE 12.2. THIRD GEODETIC FOOT OF LAGASH AND THE GREAT PYRAMID OF GIZA.

Next, we are in for even more surprises. The Octopus *talent* visits England.

It would appear that the Octopus *talent* of 29,000 grams found by Sir Arthur Evans in 1901 was the basis of the Etruscan measures of volume and weight. The Etruscan wool pound of 453.074 grams, or 6,992 grains, became the French wool pound used by British sheepherders. Queen Elizabeth I, selected this pound in the sixteenth century as a prototype for the 7,000-grain British Imperial Pound. Table 12.3 shows the evolution of the Octopus *talent*.

Name	Ratio	Volume (ml)	Weight (gm)	Comments
Amphora	64	29,086	**29,000**	307.535-millimeter cube of water
Gallon	8	3,635.7	3,625	no match
Pint	1	454.47	453.125	453.074 = Etruscan wool pound[18]
N/A	N/A	N/A	453.125	453.074 = French wool pound[18]
N/A	N/A	N/A	453.125	453.592 = British imperial pound[13]

TABLE 12.3. THE EVOLUTION OF THE OCTOPUS TALENT INTO THE BRITISH IMPERIAL POUND.

CHAPTER 13

PENDULUM 9 AND THE GEODETIC FOOT OF ATHENS

THE MYSTERIOUS PRECISION IN THE CONSTRUCTION OF THE PARTHENON

The Parthenon in Athens, Greece, was accurately measured by Stuart in 1750 and later by Penrose in 1888.[6] The dimension of the width of the Parthenon at 30.861 meters appears to be almost exactly one average arc second on the polar circumference of the Earth, 30.870 meters.[21] The small 9-millimeter error out of 30,870 millimeters was surprising considering that this true measure of the Earth was obtained in 1984 with satellite data. This level of accuracy was just not possible in 600 BCE.

FIGURE 13.1. THE PARTHENON OF ANCIENT GREECE.

Its width was exactly one arc second on the polar circumference of the Earth. The accuracy with which the attic *foot* predicts the polar circumference of the Earth has perplexed scholars for 150 years. This extreme accuracy was simply the product of luck. Pendulum 8, when timed using Venus rather than the Sun, lengthened about 0.37 percent at the latitude of Athens, eliminating almost all error.

In table 13.1, we show that the third geodetic foot of Lagash was not only used to design the Great Pyramid at Giza but traveled north to Knossos and on to Athens, picking up on the way the Minoan method of timing using the planet Venus at apogee.

Pendulum P9	Length L (mm)	*foot* 0.36 L	*talent* (gm)	Description and Latitude Degrees North
g = 9.80665	993.621	357.704	N/A	One-second beat period, standard gravity
g = 9795322	992.473	357.290	N/A	One-second beat in Babylon at 32.5352 degrees north
g = 9.797598	992.704	357.373	N/A	One-second beat in Knossos at 35.2985 degrees north
g = 9.799940	992.941	357.459	N/A	One-second beat in Athens at 37.9838 degrees north
P = 0.655738	853.710	307.336	28,943.7	Length of simple pendulum 8 in Knossos
P = 0.655738	853.914	307.409	28,964.3	Length of simple pendulum 8 in Athens
P = 0.656955	857.087	**308.551**	29,288.4	Length of Venus pendulum 8 in Athens (P 8 x 1.001856)
E1 = 191 ppm	N/A	**308.551**	N/A	308.610 millimeters measured value of attic *foot*
E2 = 300 ppm	N/A	**308.551**	N/A	30.8703 meter, one arc second at 37.9838 degrees

TABLE 13.1. THE OCTOPUS TALENT YIELDS THE ATTIC FOOT.

In table 13.1, the first four rows show the local gravity, pendulum length, and *foot* for a one-second pendulum, first for standard gravity and then for the gravity in Babylon, Knossos, and Athens. The fifth and sixth rows show the corrected period and lengths for both Knossos and Athens as well as including the weight of each *talent*. In row eight, we correct period, length, and weight using Venus as a clock rather than the Sun. Next, in the last two rows, we show that the calculated value for the width of the Parthenon is 191 parts per million shorter than the measured value and that the measured value is only 300 parts per million less than the average value for the arc second measured with satellite data (Earth according to WGS84). It seems the 150-year mystery is solved. Venus came to the rescue!

CHAPTER 14

GREECE, ROME, AND CONCLUDING REMARKS

The Athenians created the *stadion* of 600 attic *feet*. This length was almost exactly six arc seconds on the polar circumference of the Earth, or 0.1 of a true nautical mile.

Rome adopted the length of *this stadion as the stadia*; however, it contained 625 Roman feet,[28] establishing the length of the Roman foot at 296.296 millimeters. The Romans also created a mile of 8 stadia or 5000 Roman feet.

There were 75 Roman miles rather than 60 nautical miles per degree of latitude. This would establish the polar circumference of the Earth at 27,000 Roman miles. The Roman armies conquered a large part of the Western world and spread this very accurate measurement to millions, yet we may never know if the Romans or the conquered were aware of the accuracy with which their mile could measure the Earth.

Today, using modern satellite data, we find the circumference was eight Roman miles short, an error of only 0.02 percent. The Romans used the ratio of (25:24) in developing their new foot, which would lead to cultures throughout Europe adopting this ratio too and sometimes its reverse to previous standard feet as well. The resulting confusion and profusion throughout Europe would provide a strong impetus for reform.

CONCLUSION

In conclusion we have established five pendulum-based standard lengths from celestial observations. They were used in the development of three Sumerian, one Egyptian, and one Minoan standard of length. The Minoan standard traveled quite widely. It could be found in early England as in both the Scottish pound and the Troy pound, both values guaranteed by the Magna Carta of King John on June 15, 1215. These Minoan standards also seem to have made a side trip to Japan, establishing the Japanese *shaku* at 303 millimeters. The Japanese, a maritime nation, established their longest standard length, the *ri*, at 12,960 *shaku*, almost exactly 1/10,000 the polar circumference of the Earth.

Next, we discussed how the city of Lagash made three attempts to introduce a new standard based on the polar circumference of the Earth. The first version made its way to Scotland, where it appears to have been used to establish the long-foot of Stonehenge. The second version made its way to China, where it seems to have been used to establish the Chinese market foot of the Zhou Dynasty. It also seems to have appeared as the Fuss of Bern, Austria. The third and most accurate standard was used in the design of the Great Pyramid of Giza in Egypt and two thousand years later in the design of the Parthenon in Athens, Greece. Today, almost five thousand years later, we are surprised to find it still in use in defining the weight of the U.S. pound.

When beginning on this journey of discovery into the standards of ancient Sumerians of 3000 BCE, I had no idea that it might lead me on a journey of almost five thousand years to the present day and was quite surprised to find that both the British imperial and U.S. pounds are related to the polar circumference of the Earth through the third geodetic foot of Lagash.

I hope you enjoyed reading some of this journey through time. There is still a lifetime of adventure awaiting the patient explorer. Won't you join me in uncovering the past? Finally, can there any longer be any doubt that the pendulum was used in the development of ancient metrology?

NORMAL MINA VALUES DERIVED FROM INSCRIBED WEIGHTS

SUMERIAN NUMERATION AND METROLOGY

A THESIS
SUBMITTED TO THE FACULTY OF
THE GRADUATE SCHOOL
OF THE UNIVERSITY OF MINNESOTA
by
Marvin Adall Powell, Junior

IN PARTIAL FULFILLMENT OF THE REQUIREMENTS
FOR THE DEGREE OF
DOCTOR OF PHILOSOPHY

December, 1971

1	Mina	Key #	NOTE	Description	#	P	1
2	680.485	680.485	Mina D	Pear Shaped	46	P2/P7	2
3	607.485	101.32	10 shekel				3
4	564.1	188.05	1/6 mina				4
5	558	4.65	"baz"				5
6	557.81	184.27	1/6 mina	Bibliog	71	P3	6
7	557.81	184.27	1/6 mina	Bibliog	72	P3	7
8	549.44	52.365	3 shekel				8
9	538.5	35.9	3 shekel				9
10	538	538	ma na				10
11	537.1	5371	10 mina				11
12	534.162	178.054	1/6 mina	Duck Shalmaneser	85	P7	12
13	522	2.9	little mina	Ellipsoid Susa	229	P6	13
14	521.1	173.7	1/3 mina	Bomb Shaped	75	P6	14
15	516	17.2	2 shekel	Duck Susa	137	P1	15
16	516	43	5 shekel	Duck Bibilog	103	P1	16
17	514.8	4.29	1/4 shekel	Duck from India	212	P1	17
18	511.3	127.837	1/8 mina	Duck Bibilog	82		18
19	510	170	1/3 mina				19
20	506.6	1520	3 mina silver				20
21	505.5	168.5	zadium				21
22	505	505	1 mina				22
23	504.925	2019.7	4 mina				23
24	504.625	60555	2 talents				24
25	504.56	2522.8	5 mina				25
26	504.28	5042.805	10 mina ?				26
27	502.018	15060.552	30 mina				27
28	502.195	2510.975	5 mina				28
29	500.172	166.724	1/3 mina				29
30	499.29	665.729	3/2 mina	Bronze Lion	47	P7	30
31	498.67	4986.712	10 mina	Stone Duck	18	P7	31
32	498.468	41.539	5 shekel	Ellipsoid from Telloh	108	P7	32
33	498.024	1992.099	4 mina	Bronze Lion	30	P7	33
34	498	41.5	5 shekel	Stone Duck	18	P7	34
35	497.741	14933.757	15 mina				35
36	497.5	9950	20 mina				36
37	497.5	497.5	Dudu 1 mina				37
38	497.5	995	Sulgi 2 mina				38

TABLE A 1-37

#	Mina	Key #	Note	Description	#	P	
37	497.5	497.5	**Dudu 1 Mina**				37
38	497.5	995.0	Sulgi 2 mina				38
39	496.40	198.4	1/5 mina				39
40	496	248	Sulgi 1/2 mina				40
41	496	248	1/2 mina				41
42	495.102	82.517	10 shekel				42
43	494.6	29680	1 Talent				43
44	493.32	123.33	15 shekel				44
45	493	4930	10 mina				45
46	492.9	164.3	1/3 mina				46
47	491.53	4915.3	2.5 mina				47
48	491.38	245.69	1/2 mina				48
49	491.22	4930	Zadin				49
50	489.154	978.3	**Mina N**	Pyramid Berriman	93	P2	50
51	489.6	244.8	1/2 mina	Ellipsoid	64		51
52	487.2	40.6	5 shekel	Ellipsoid	114	P2	52
53	486.3	14589.6	30mina	Alabaster Duck	11	P2	53
54	486	40.5	5 shekel	Truncated Spindle	116		54
55	448.40	2422.0	5 mina				55
56	481.07	240.5	1/2 mina	**Bronze Lion**	65		56
57	480.145	480.145	1 mina	Bronze Lion	53	P3	57
58	480	40	5 shekel	Duck Susa	119	P3	58
59	480 `	40	5 shekel	Ellipsoid	120	P3	59
60	479.85	159.95	1/3 mina	Elongated Barrel	81	P3	60
61	479.6	39.363	5 shekel	Duck Susa	121	P3	61
62	477.438	2864.629	3 mina	Bronze Lion	23		62
63	477.28	954.566	2 mina	Bronze Lion	43		63
64	477.2	119.3	15 shekel				64
65	477	5.3	2/3 shekel	Elongatd Barrel	81	P2	65
66	476.1	79.35		Elipsoid Biblioc	92		66
67	475	1425.0	3 mina	Barrel Shape	55	P3	67
68	473.36	236.678	1/2 mina				68
69	473.23	946.462	2 mina				69
70	472.8	1931.229	4 mina				70
71	468.39	468.388	1 mina	Bronze Lion	54		71
72	465.004	232.502	1/2 mina	Duck Louvre	67	P5	72
73	456	0.95	22.5 SSE	Duck Louvre	256		73
74	453.467	680.485	2/3 Mina D	Pear Shaped	255	P2	74

TABLE B 37-74

TALENT WEIGHTS

Corolla Numismatica, Numismatic Essays
in Honor of Barclay V. Head
Minoan Weights and Mediums of Currency
from Crete, Mycenae, and Cyprus
By Sir Arthur Evans, pages 342 and 358

Found in the Fifteenth West Magazine of the Palace at Knossos in 1901, we see below on the right a magnificent gypsum talent weight with octopus relief. It is 42 centimeters high and weighs almost exactly 29,000 grams (discussed on page 342).

FIGURE 12.1. OCTOPUS TALENT WEIGHT.

But in 1903, a still more fortunate discovery was made by the Italian Mission, under the direction of Professor Halbherr, as shown in table 1. In the course of the excavations of the small Palace or Royal Villa of Hagia Triada, there were discovered hordes of bronze ingots, nineteen in all. It is noteworthy that two examples weigh exactly 29,000 grams, the precise amount scaled by the standard palace weight from Knossos, with the octopus reliefs. This, as already pointed out, represents a light Babylonian *talent*.

Another series of seventeen similar ingots, besides fragments, was found in the sea. They are now in the museum at Athens. These had unfortunately been much corroded by the sea, which may deduct from their metrological value. Professor Pigorini gives their weights as in table 2.

#	Weight (gm)	P		#	Weight (gm)	P
1	**29400**	**P2**		11	27300	
2	29500			12	29500	
3	**29400**	**P2**		13	27000	
4	29400			14	**29300**	**P9**
5	29900			15	**32000**	**P7**
6	30700			16	29200	
7	**27900**	**P5**		17	**29000**	**P8**
8	**29400**	**P2**		18	**29000**	**P8**
9	30000			19	27600	
10	**30900**	**P1**				

TABLE 1. THE HAGIA TRIADA INGOTS BY JOSEPH HAZZIDAK.

#	Weights (gm)	Fraction of Talent		#	Weights (gm)	Fraction of talent
1	**13320**	x 2 = 26640		10	13230	x 2 = 26640
2	**13860**	x 2 = 27720		11	17640	x 2 = 35280
3	**17000**	x 2 = 34000		12	5320	x 5 = 26600
4	**17000**	x 2 = 34000		13	11650	x 3 = 34950
5	**11970**	x 3 = 35910		14	6930	x 5 = 34650
6	**13320**	x 2 = 26640		15	9450	x 3 = 28350
7	13260	x 2 = 27240		16	12900	x 2= 25800
8	12600	x 2 = 25200		17	1008	x 30 = 30240
9	11340	x 3 = 34020				

TABLE 2. THE CHALKIS WEIGHTS FOUND BY PROFESSOR PIGORINI.

APPENDIX 3

BERRIMAN PHOTOS, WEIGHTS, AND LENGTHS

A. E. Berriman O.B.E in *Historical Metrology* (first published in 1953) provides several examples of standards of length, volume, and weight. While we may not agree with all his conclusions, this book contains very valuable information for the student of ancient metrology.

Some of the data and photographs from this book, which are used in our report, are listed below:

Fig 4.1 (Page 4) Statue of "Gudea's Rule," circa 2175 BCE, The Louvre

FIGURE 4.1 GUDEA'S RULE.

Fig 4.2 (Page 64) Entemena's Vase, Lagash, now in the Louvre

FIGURE 4.2. ENTEMENA'S VASE.

Fig 5.1 (Page 57) *mina* N Third dynasty of Ur, British Museum, Exhibit 91005

FIGURE 5.1 *MINA* N.

Fig 5.3 (Page 56) *mina* D of Lagash, Ashmolean Museum, Exhibit 1921.870

FIGURE 5.3. *MINA* D.

Fig 5.2 (Page 62) Babylonian limestone duck, Ashmolean Museum, Exhibit 1912.1162

FIGURE 5.2. LIMESTONE DUCK.

Fig 12.1 (Page 135) Octopus *talent* weight, A copy is in the Ashmolean Museum

FIGURE 12.1. OCTOPUS WEIGHT.

LENGTHS AND ANGLES

Page 29 Length of Chinese ch ih shown in a table of linear units

Page 79 Cotangent of angle Great Pyramid = pi/4, measured by Petrie

Page 122 Length of Roman foot, Stettinius Aper's monument in Rome

DRAWING OR SKETCH

Page 90 Petrie Collection, University College, London

Egyptian bronze bowls number eight and twenty-seven

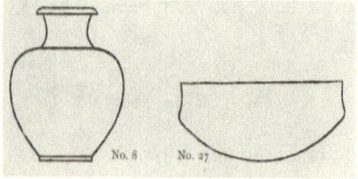

PENDULUM MATHEMATICAL ANALYSIS

The period of a physical pendulum with corrections for the angle of swing, the weight of string, and both weight and diameter of ball.

(1) Increasing the angles of swing from a small value to a practical one

$(P/P0) = 1 + (1/16) \, a^2 + (11/3072) \, a^4 + (173/73280) \, a^6$

Where a is the maximum angle of swing to each side in Radians.

For example, using a practical maximum swing of 0.1 pendulum lengths

a = 5.74 degrees, or 0.1001819 Radians yields an increased period of $(P\,/P0) =$

$1 + 0.000625 + 0.000000358 + 0.00000000002346 = 1.000625358$

(2) Increasing mass of a real string from the ZERO mass assumed in the case of a simple pendulum will reduce the period as follows:

$(T/P0)^2 = (M/m + 1/2)/(M/m + 1)$

Where M/m is ratio of pendulum mass to string mass

For example, a 60-gram steel ball used with a 0.60-gram waxed flax string yields $(T/P0)^2 = (100 + 1/2)/(100 + 1) = 0.99505$

Now taking the square root, we find $(P/P0) = 0.997522$

(3) Increasing the size of the ball from a point mass will increase the inertia of the ball and the string tending to increase the period as follows:

$(P/P0)^2 = (1 + 2/5 \ (R/L)^2 + 1/3 \ (m/M)$

Where R is the radius of the ball and L is the length of the string. For example, a 60-gram ball one inch in diameter with a one-meter string yields $(T/Po)^2 = 1.0033978$

$(1 + 2/5 \ (0.0127)^2 + 1/3(0.01) = 1 + 0.0000645 + 0.00333 = 1.0033978$

Now taking the square root, we find $(P/P0) = 1.001697$

Now combining the effect of the properties of mass yields

$(P/P0) = 0.997522 \times 1.0033978 = 0.999215$

Now adding the effect of the swing angle

$(P/P0) = 0.999215 \times 1.000625358 = 0.999840$

A few sample calculation serves to illustrate this effect:

Swing, Fingers	String Weight	Ball/String Ratio	Length Correction
3	173.7 milligrams	439.6	0.000000
6	614.2 milligrams	124.3	0.000000
10	1.711 grams	44.62	0.000000
15	4.113 grams	18.57	0.000000
20	8.289 grams	9.213	0.000000

TABLE 1. THE WEIGHT OF STRING REQUIRED OBTAINS THE THEORETICAL PERIOD FOR A 60-FINGER (1-METER) PENDULUM WITH A 1-FINGER-DIAMETER 76.3-GRAM COPPER BALL FOR A NUMBER OF SWING DISTANCES.

Next, we computed the string weight and ball/string ratio required to completely cancel the effect of swing distance with a ball 2 fingers in diameter weighing 172.6 grams.

Swing, Fingers	String Weight	Ball/String Ratio	Length Correction
3	441.5 milligrams	390.9	0.000000
6	1.437 grams	120.1	0.000000
10	3.920 grams	44.03	0.000000
15	3.920 grams	44.03	0.000000
20	3.920 grams	44.03	0.000000

TABLE 2. THE WEIGHT OF STRING REQUIRED OBTAINS A THEORETICAL PERIOD FOR A ONE-METER PENDULUM WITH A TWO-FINGER (33.33-MILLIMETER) DIAMETER 172.6-GRAM COPPER BALL FOR A NUMBER OF SWING DISTANCES.

GRAVITY AS A FUNCTION OF LATITUDE

File: GRACE globe animation.gif From Wikipedia, the free encyclopedia

The mathematical model

If the terrain is at sea level, we can estimate g:

$$g_\phi = 9.780327 \left(1 + 0.0053024 \sin^2 \phi - 0.0000058 \sin^2 2\phi\right)$$

The first three terms in the International Gravity Formula of 1967 for calculating the acceleration of gravity as a function of latitude

City	Latitude Degree	Sine X	Sine 2X	Gravity
Lagash	31.4068	0.521111	0.889525	9.79487
Ur	30.9575	0.514402	0.88225	9.794032
Babylon	32.5352	0.53782	0.906826	9.795322
Luxor	25.6872	0.433458	0.781242	9.790064
Memphis	29.8431	0.497631	0.863274	9.793163
Knossos	35.2985	0.577836	0.943205	9.797598
Athens	37.9838	0.618439	0.970159	9.79994
Paris	48.8566	0.753065	0.990952	9.80968
London	51.5074	0.782688	0.974312	9.81209
New York	40.7128	0.652268	0.988823	9.802333
Los Angeles	34.0522	0.559948	0.927865	9.796539

Data File #1 Gravity at selected Latitudes

Standard Gravity = 9.780327 meters per second squared

SOLAR TIMING

Establishing the interval of time for the Sun to travel one diameter in the sky

One solar day = 86,400 seconds

The Sun's angular velocity = 240 seconds/degree

The Sun's angular velocity = 0.004167 degrees/second

The Sun's apparent diameter = 1919.62 arc seconds

The Sun's apparent diameter = 0.533228 degrees

The Sun's apparent radius = 0.266614 degrees

Viewing the Sun's image created by a pinhole in the temple roof

Below, see examples of the Sun image as a function of height

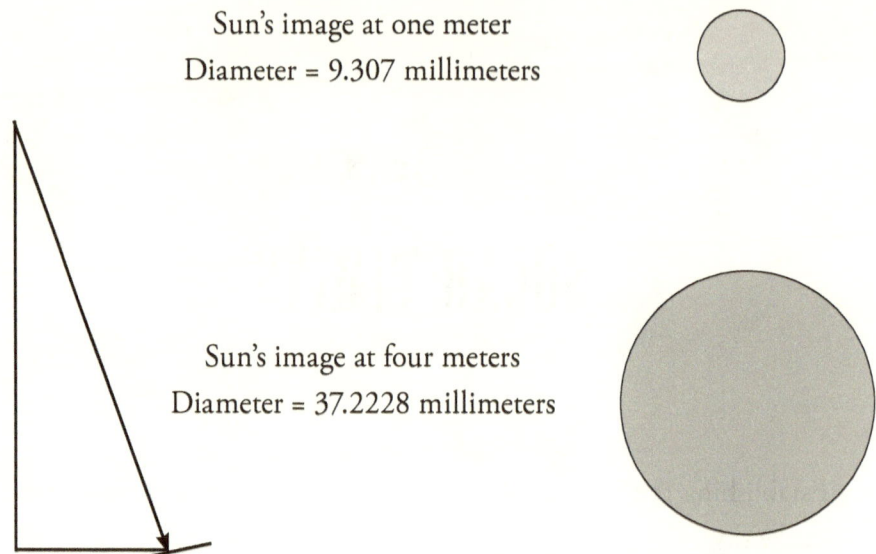

Sun's image at one meter
Diameter = 9.307 millimeters

Sun's image at four meters
Diameter = 37.2228 millimeters

Transit time of the Sun for one diameter = 127.975 seconds

Transit time of the Sun for one degree = 240 seconds

The Sun's image would be quite bright shining from a pinhole in the ceiling of a darkened room. For truly accurate results, the Sun should be timed for a full twenty-four hours. Anything less than timing for thirty degrees or two hours would be insufficient.

ESTABLISHING THE INTERVAL OF TIME FOR THE MOON

To travel one diameter in the night sky:

Earth sidereal day = 86164.08 seconds

Earth angular velocity = 239.3447 seconds/degree

Earth angular velocity = 0.004178075 degree/second

The lunar period = 27.321661 days = 2,360,591.5 seconds

Lunar mean angular velocity = 6557.20 seconds per degree

Lunar mean angular velocity = 0.000152504 degree/second

The mean lunar orbit eccentricity = 0.0549

Applying the equal-area rule:

The orbital velocity at perigee = 1.0549 x the mean = 0.000160877 degree/second

The orbital velocity at apogee = 1/1.0549 x the mean = 0.000144567 degree/second

For an observer on Earth:

The Moon's mean apparent rotation = Earth rotation - Moon mean rotation

The Moon's mean apparent rotation = 0.004025571 degree/second

The Moon's mean apparent rotation = 248.412 seconds/degree

The Moon's apparent rotation at perigee = 0.004017198 degree/second

The Moon's apparent rotation at perigee = 248.930 seconds/degree

The Moon's apparent rotation at apogee = 0.004033508 degree/second

The Moon's apparent rotation at apogee = 247.923 second/degree

The mean diameter of the Moon = 31.12 arc minutes = 0.518667 degree

The diameter of the Moon at perigee = 33.40 arc minutes = 0.556667 degree

The diameter of the Moon at apogee = 29.33 arc minutes = 0.488833 degree

The mean transit time of the diameter of the Moon = 128.843 seconds

The transit time of the diameter of the Moon at perigee = 138.571 seconds

The transit time of the diameter of the Moon at apogee = 121.193 seconds

REFERENCES

1   Ronald E. Zupko, *British Weights and Measures* (The University of Wisconsin Press), 20. Table 1 Description, Anglo Saxon and British Furlong.

2   E. Janhke, and F. Emde. *Table of Functions*, 4th ed. (New York: Dover Publications, 1945), 85. Table V. Complete Elliptical Integrals (Mathematics of the Pendulum).

3   Arthur Bronwell, *Advanced Mathematics in Physics and Engineering* (New York: McGraw-Hill, 1953), 137–139.

4   Glen Thorncroft, "Accuracy of Gravity," http://www.calpoly.edu/~gthorncr/ ME302/documents/AccuracyofGravity.pdf.
    Gravity Equations, figure 1 and table 1.

5   A. E. Berriman, *Historical Metrology* (E. P. Dutton & Co.,1953), 53–54.
    Gudea's rule, figure 4A, table 4A, line 9.

6   Berriman, *Historical Metrology* (E. P. Dutton & Co.,1953), 63–64.
    Pg. 5: Mina N, figure 5A, table 5C, line 2
    Pg. 47: Chinese market foot (318-millimeter), table 11A, line 6
    Zhou Royal Ch'ih, table 5D, line 3
    Pg. 56: Mina D, figure 5C
    Pg. 62: Limestone Duck, figure 5B
    Pg. 63–64: Entemena's Vase, figure 4b, table 4a, line 12
    Pg. 90: Figure7.1 Egyptian bowls eight and twenty-seven, volume ratios, Table 7.1; Foot, finger, and royal cubit from bowl eight, Table 7.2, lines 4, 5.7, 8, 9; Foot, finger, and royal cubit from bowl twenty-seven, Table 7.3, lines 3, 4, 5, 6
    Pg. 116: Attic stade = 600 attic feet = 625 Roman feet

7   Arthur J. Evans, "Minoan weights and mediums of currency from Crete, Mycenae, and Cyprus," *Corolla Numisimatica, Numismatic Essays*, 358.
    Pg. 342: Octopus talent, figure 12A, table 12a, line 10
    Pg. 358: Talent #10, 30,900 grams, table 4b, line 1; Talent #13 (4), table 8A line 8, table 8A, line 7, table 8B line 1; Talent #15 (14), table 11B line1

8   M. A. Powell, *Sumerian Numeration and Metrology* (University of Michigan, 1973), 205.
    Weights table 4A line 4, 5, 6, table 4b line 3 and 4
    Weights table 5A line 4, 5, 6, table 5b line 3, 4, 5
    Weights table 5C line 2, 3
    Weights table 6A line 4, 5, 6, 7, 8, table 6b line 7, 8
    Weights table 8A line 4, 5, 6, 7, 8, table 8b line 4, 5
    Weights table 10 line 7, 8
    Weights table 11B line 3, 4,5, 8, 9

9   P. Guilhiermoz, "A propos d'une publication récente, 'Pied de Terre' of Bordeaux, France," table 5d, line 4.

10 Carlo Zaccagnini, *The Assyrian Lion Weights from Nimrud" and the "Mina of the Land" "Michael"* (Tel Aviv: The Archeological Center Publications, Jaffa, 1999).
    Assyrian Lion Weights, figure 6A, table 6A lines 11, 12, table 6B line 9
    Assyrian Lion Weights, table, 8b line 5
    Assyrian Lion Weights, table 11b, lines 7–8

11 Wikipedia, "Ancient Egyptian units of measurement."
    Egyptian New and Old Kingdom weights, table 7D.

12 Wikimedia commons, "Beautiful photo of the Great Pyramid at Giza."

13 (26) Zopko, *British Weights and Measures*, 155.
    Pg. 155: English mercantile pound (6,750 grains), table 8c, line 2
    Pg. 155 English Troy Pound (5,760 grains), table 8c, line 6
    Pg. 156: British imperial pound (7,000 grains), table 13A, line 5

14 (25) *Full Sail Online*, "Numbers, counting, and dates," sci.lang.japan FAQ.
    Japanese shaku = 303.6 millimeters
    Japanese ri = 12,960 shaku = 1/10,000 circumference of the earth

15 WGS 84b 1984 "every degree."
    The polar circumference of the Earth.

16 *The British Journal for the History of Mathematics.*
    The Stonehenge long-foot, 322.9 millimeters

17 (12) Everything.explained.today A-Z Contents O OB OBS Select obsolete
    German units of measurement.
    Steinbrecherfuss of Bern Austria, table 11A line 7

18 Bee Wilson, *Swindled* (Princeton University Press, 2008).
    65.Etruscan wool pound, table 12A line 3
    French wool pound, table 12A line 4

19 (14) Sir W. M. F Petrie, *Inductive Metrology* (London: H. Saunders, 1877),
    section 21–39 figure 13C, perimeter of the Great Pyramid.

20 13 WGS 84b 1984 "every degree."
    The polar circumference of the Earth, table 13b, line 8
    The average value of one arc minute, table 13a, line10
    The value of one arc minute in Athens, table 13a, line 11

www.ingramcontent.com/pod-product-compliance
Lightning Source LLC
Chambersburg PA
CBHW022103170526
45157CB00004B/1455